On Spin, Mass, and Charge

Greg Feild

November 29, 2016

Abstract:

This paper continues the investigations into a simpler standard model last visited in "On Gravitation and Electric Charge".

In this study, we find the massive gauge bosons (W+,W-,Z) are unnecessary to explain the weak interaction.

We conclude there is one massless gauge boson responsible for all particle interactions; the 'photon'.

In addition we offer the usual ruminations on aspects of this new standard model important for cosmology and provide an overall summary of our new quantum mechanical theory of gravitational interactions.

Preface:

Dear reader(s),

This book is a culmination of our investigations into particles and their interactions; the thoughts we have developed together over the course of three previous papers on integrating gravity into the standard model.

I believe the result is a clear, concise and coherent picture of all elementary particle interactions, as well as all gravitational interactions.

These papers also include some more speculative musings on the fundamental nature of elementary particles, space, and time!

Obviously, there is still a LOT of work to be done, so please join the effort!

It may not bend your mind, but it's still physics ... and fun!

Greg F.

Introduction:

In this fourth book on 'grand unification' and a simpler standard model, we continue the investigations begun in our three previous books on the quantum mechanical nature of gravitational interactions (1,2,3).

Previously, we found we could do without quarks, gluons, the color charges and the weak charge.

In this paper, we find we can dispose of the massive electroweak exchange bosons as well.

Our conclusion is there is one massless gauge boson that interacts with all particles, coupling to the particle's total relativistic mass-energy/charge.

We will begin with the proverbial 'matchbook summary' of our new, more modest standard model ("the universal model", if you will) complete with the current standard model of weak interactions removed.

In later sections we will try to justify the demotion of the W and Z bosons as force carriers.

In addition, we will examine several slightly suspect symmetries and conservation laws in light of our new universal model.

We also offer a qualitative discussion on how the universal model would describe cosmological phenomena; in particular, large scale intergalactical interactions.

The universal model:

In the universal model, all particle interactions are mediated by the exchange of a single massless, spin 1 boson, heretofore and henceforth to be known as the photon.

The photon couples to a particle's 'total coupling charge', tcc, which is defined as

$$tcc = m(1+e') \tag{1}$$

where m is the particle's relativistic mass-energy and e' is the particle's charge to mass ratio

$$e' = e/m\_rest \tag{2}$$

and e is the electric charge and m_rest is the rest mass of the particle.

Classically, the complete, relativistic Lorentz force on a particle due to a specified distribution of mass and charge is now given by

$$F = (me')*E + (me')*(vxB) + m*F\_g + m*(vxB\_g) \tag{3}$$

where F_g is calculated in the usual way, and B_g has been derived in a previous paper (3) and in just the way one would expect.

For quantum mechanical calculations, the propagator for an interaction between any two particles is

$$f(q) = (tcc\_1)*(tcc\_2)/q^2 \tag{4}$$

where q is the four-momentum of the exchanged photon. We have left out any multiplicative factors such as G, alpha, etc., as it is not yet clear what final role they will play. Certainly, we will defer to the experts in such matters.

Finally, the photon has a coupling charge, m_g, given by

$$m\_g = hbar*nu/c^2 \tag{5}$$

where nu is the frequency of the photon and c is the speed of light.

This is our new standard model in a nutshell (if you'll pardon the mixed metaphor!).

Building blocks of matter:

In the universal model, there are two fundamental fermions which comprise all ordinary matter; the electron and the electron neutrino.

The electron neutrino is considered to be a massive, point particle with spin ½ and a magnetic moment hypothesized to be

$$\mu_v = e \hbar / 2 m_v c \qquad (6)$$

Here, e is no longer considered to be the 'electric charge' *per se*, but rather a dimensionless proportionality factor, as we shall see next.

The electron mass has previously (1) been shown to be

$$m_e = e \cdot m_v \qquad (7)$$

where e is the electric charge and m_v is the neutrino mass.

The electron magnetic moment can be now be written in terms of e and the magnetic moment of the neutrino

$$\mu_e = (1/e) \cdot \mu_v \qquad (8)$$

In this model, the electron is 'just' the neutrino with an electric charge. The electron has increased in mass-energy and coupling charge by a factor of e relative to the neutrino.

However, the electron's total electromagnetic energy remains constant relative to the neutrino in view of equations (7) and (8) and the electron's potential interaction with an external field.

Previously, we have suggested that a quantum mechanical combination of mass and spin gives rise to the 'electric charge'. In addition, the direction of the electron's spin would determine the sign of its electric charge.

In this manner, perhaps, we can can claim gravity is both attractive and repulsive as mass is ultimately responsible for electric charge. In addition, in our revised model, mass currents traveling in opposite directions will give rise to oppositely signed magnetic fields as can be seen from equation (3).

However, it seems there is still a slight asymmetry in the overall attractiveness of this 'universal force' as can be seen from equation (1). Macroscopically, this appears to be true.

(Let us say for fun, that this 'asymmetry' drives the eternal death and rebirth of our universe, and is the 'reason' we are all here! Why not?)

Microscopically, quantum mechanics and spin come to the rescue. Imagine two identical neutrinos under mutual gravitational attraction. It seems surely they will 'collide' or come into actual "physical contact". This can not happen however, due to the interaction between the magnetic moments of the two neutrinos.

Dimensionality of e:

We have already seen in equations (1), (6), and (7) that the electric charge e functions as a dimensionless factor in our new universal model. Its new role seems to be in the quantization of particle mass!

(It also seems this new role for "e" will be no problem for field theory, but I'm not so sure about trying to put it retroactively into classical electrodynamics ...)

In reference (1), we derived a formula for the fine structure constant, alpha, in terms of the electric charge, e, the electron mass, m_e, and the proton mass, m_p

$$\text{alpha} = (m\_e^2/m\_p^2)*e^2 \qquad (9)$$

As alpha is dimensionless by design, equation (9) also implies that e should be dimensionless, that is, if the following relation

$$m\_e/m\_p \sim= 1/(4*pi*epsilon\_0*hbar*c)^{1/2} \qquad (10)$$

can be shown to be true in some 'unit space'.

Weak interactions:

There have been indications since we first began these investigations and determined that a particle's 'weak charge' was equivalent to its mass, that this tiny coupling charge may be enough to explain the 'weakness' of the weak interaction. Still, we have kept the W and Z until now out of convenience and expediency (and with no other clear ideas on the matter). In this section, we will examine whether the W and Z are really necessary and whether we can explain them away!

Let's begin historically, with beta decay; whereby a neutron turns into a proton;

n ---> p + e- + v_e^bar   (11)

In the standard model, beta decay occurs when a down quark emits a massive W- boson to become an up quark. The W- particle then decays into an electron and an electron antineutrino.

(u,d,d) ---> (u,u,d) + e- + v_e^bar   (12)

This process is highly suppressed due the massiveness of the W and not due to the nature of the weak charge.

In our new model, where partons are leptons, beta decay now looks like this

(e+,e-,v_e^bar) ---> (e+,e-,e+) + e- + v_e^bar   (13)

Here, the antineutrino emits a photon which then decays into an electron and a positron. In terms of the Feynman diagram, the coupling charge at the antineutrino vertex is the neutrino mass (which 'suppresses' the interaction). The coupling charge at the electron vertex is the tcc of the electron.

This view of the weak interaction does not require a charged quantum mechanical force field. It is already a 'stretch' to imagine fields possessing energy and momentum, much less mass and electric charge!

As a second example, let's look at muon decay,

mu+ ---> v_mu + e+ + v_e   (14)

Again, in the standard model, the muon emits a charged W particle turning into a muon neutrino. The W+ then decays into a positron and an electron neutrino.

In addition to requiring a virtual field to carry the mass and electric charge away (as in beta decay), it just seems unnatural that a particle should change into another particle, particularly in such an *ad hoc* way.

In our new model, it is tempting to try the "tree level diagram" we used to explain beta decay. Here, we would have the muon emitting a photon to become an electron. The photon would then decay into a neutrino antineutrino pair of one's choosing.

This is tempting as it would be a direct decay of a moun into an electron via photon emission, as if, perhaps, the muon were an excited electron.

However, it is less appealing as it violates conservation of lepton number, although this quantity is not thought to be absolutely conserved. Since technically we have moved beyond the standard model, this channel may be open to the muon and perhaps could be worth a closer look.

In our new model, in order to conserve lepton number, the muon decay would have to proceed via a second order "box" diagram with four internal virtual photons. Of course, this same box diagram also allows for neutrino mixing!

So, we have accounted for the weak interaction, *within the standard model*, but without having to invoke charged, massive virtual exchange bosons.

This new weak theory is QED with mass as the coupling charge.

"Hold on", you might say, "haven't the W and Z been observed in particle collisions?"

In our new theory, these particles represent temporary resonant bound gravitational states of the colliding leptons. (For W production in ppbar collisions, it is equally likely to imagine the event as deep inelastic scattering between an electron and a neutrino.)

Our conclusion is the electroweak theory gives the correct energy scale for the unification of electromagnetism and the 'weak interaction', but an incorrect picture of the mechanism.

As an example, let's consider Z production in electron positron collisions.

In the standard model, the electron and positron annihilate producing a Z (or technically Z/gamma!) which can then decay into a pair of leptons or a pair of quarks which then decay into 'jets' of particles. So, our new model must explain this hadronization.

In the universal model, the electron and positron can annihilate to a photon or form a temporary, gravitationally bound, mesonic state.

Once the electron and positron are bound gravitationally, gravity effectively starts to behave like QCD in terms of confinement. The particles can only 'break apart' by forming gravitational bonds with neighboring particles.

Q.E.D.!

Symmetries and conservation laws:

A simpler standard model should have simpler symmetries!

We began in reference (1) by addressing the apparent asymmetry in nature between matter and antimatter. There, we hypothesized the world is composed of equal parts matter and antimatter; the antimatter being bound inside protons and neutrons. In our new model, partons are leptons (and anti-leptons), and there are no more quarks.

No more quarks means no more isospin or strangeness conservation.

We have also seen earlier in this paper that our new model may encourage violation of lepton number conservation as well.

Cosmology:

We claim our new universal theory can explain all elementary particle interactions as well as large scale cosmological phenomena.

For quantum mechanical problems there is equation (4), but for describing the interactions of large scale bodies we have the Lorentz force of equation (3) and Newton's three laws of motion.

For neutral bodies such as galaxies and stars, etc., we can use the 'cosmological Lorentz force'

$$F = m^*(F\_g) + m^*(v \times B\_g) \qquad (15)$$

where, of course, m is the relativistic mass.

One can see there is now the prospect for repulsion between two bodies moving in opposite directions. In addition, a spinning galaxy now creates a 'gravitational magnetic field' and can be assigned a series of magnetic moments.

Obviously, if equation (15) is correct, then all the derivations already done in the study of electrodynamics can be applied directly to cosmology!

Build a particle workshop:

In this section we attempt to define an elementary particle.

Appealing to the notion of the photon as the most elementary elementary particle; we conclude a particle is spin (with energy).

Spin is an invariant quantity and an inherent property of a particle. You cannot transform quantum mechanical spin away by changing reference frames.

Let's list our three elementary particles in order of increasing 'complexity'.

A photon is "a spin of 1 with energy hbar*nu".
A neutrino is "a spin of ½ with mass-energy m_v".
An electron is "a spin of ½ with energy m_e, and charge e".

We've already hypothesized how we might 'generate' a massive, electrically charged particle from a massive neutral charged particle (3). Otherwise, this progression might seem obvious and banal.

However, even if we take spin to be the most fundamental expression of a physical particle, it is unclear how energy would 'attach' to it.

We know the energy of a photon is proportional to its frequency. However, when a photon loses or gains energy it does not change in size or mass and its spin remains a constant.

The only physical attribute the photon possesses is spin. This spin can be positive or negative relative to the transverse direction of the photon (the classical analogue being circular polarization of light).

We hypothesize that the frequency, nu, of a free photon is the rate at which the photon spin flips, changing polarization harmonically as it barrels toward its target at the speed of light. This 'rate of spin flip' would be what would give the photon a particular energy and momentum.

We note, the photon carries just enough quantities to perform its job, which is transferring energy and momentum (and spin) between interacting fermions!

Massive particles with magnetic moments cannot perform such spin flips, so they take on energy in the form of extra mass. If we assume the spin and the magnetic moment are constants of the motion, then if a particle increases in mass, it must increase in 'size'.

Numerology:

We have previously confessed our ineptitude with 'natural units' and thrown off the constraints of dimensional analysis in our current reappropriation of the constant formerly known as the electric charge, e.

Now, we make the following observation about the neutrino mass m_v

$$m_v \;\sim\sim= \; \hbar/2 \qquad (16)$$

ignoring units *and* orders of magnitude!

The electron neutrino can be thought of as the smallest possible bit of matter with an angular momentum and mass equivalent to a 'spin ½' taking on one "quantum of action". It then follows from equation (7) that the mass of the electron is

$$m_e = e \cdot \hbar/2 \qquad (17)$$

We can also derive these fundamental relations for the electron neutrino and electron magnetic moments,

$$\mu_v = e/c \qquad (18)$$

and

$$\mu_e = 1/c \qquad (19)$$

in some crazy (universal) units!

Conclusion:

We began these investigations with the goal of incorporating gravity into the standard model of the three other fundamental forces of nature.

As we progressed, it became apparent the four forces were quite similar under the assumption that a particle's coupling charge was proportional to its mass.

Finally, we we had to conclude there is one fundamental force; gravity!

References: Books by Greg Feild

1. "A quantum mechanical theory of gravitational interactions", CreateSpace Independent Publishing, August 29, 2016

2. "Observations on the quantum mechanical nature of gravity", CreateSpace Independent Publishing, October 8, 2016

3. "On gravitation and electric charge", CreateSpace Independent Publishing, November 1, 2016

Notes:

www.ingramcontent.com/pod-product-compliance
Lightning Source LLC
Chambersburg PA
CBHW070722210526
45170CB00022B/1709